THE ULTIMATE GUIDE TO LAUNCHING A MODERN BUSINESS PHONE SYSTEM EVEN IF YOU'RE NOT TECH SAVVY

David Bridge

Copyright © 2016 **David Bridge**
.

All rights reserved. No part of this book may be reproduced or transmitted in any form or by any means, electronic or mechanical, including photocopying, recording or by any information storage and retrieval system without written permission of the publisher, except for the inclusion of brief quotations in a review.

Table of Contents

INTRODUCTION 1

CHAPTER 1 - BIGGEST MISTAKES LAUNCHING A NEW MODERN SYSTEM 3

WHAT'S THE BIGGEST MISTAKE YOUR PROSPECTS MAKE? 3
NOT ASKING ENOUGH QUESTIONS AHEAD OF TIME. 4
NOT TAKING A LOOK AT YOUR ENTIRE PLATFORM. 5
NOT GETTING EASY TO USE PHONES. 6
NOT UTILIZING THE FULL CAPABILITY OF THE SYSTEM. 7

CHAPTER 2 – THE TELECOMMUNICATIONS SPECTRUM AND THE BEST SYSTEM. 9

CHAPTER 3 – CALL HANDLING PHILOSOPHIES AND STRATEGIES. 14

HOW COULD THEY SCREEN THEIR CALLS? 15
WHAT'S THE BIGGEST MISTAKE A RECEPTIONIST MAKES SCREENING CALLS? 18
ANOTHER CALL-HANDLING STRATEGY QUESTION IS; HOW DO YOU HANDLE CALLS FOR A SPECIFIC DEPARTMENT LIKE SALES OR CUSTOMER SUPPORT? 19

CHAPTER 4 – CONNECTING MULTIPLE LOCATIONS 22

SO I GUESS IT DOESN'T MATTER HOW FAR AWAY THEY ARE? 23
SO SHOULD I IMPROVE MY INTERNET CONNECTION? 24
WHAT IF ONE PERSON WORKS IN MORE THAN ONE LOCATION? ...24

How do I keep from tying up multiple lines connecting another location? 26
Will I know they're on the phone? 27
What if I want an intercom in my offsite office? 28
What is fiber ethernet? 28
Can set it up to ring a different site, program it to ring a different site? 29

CHAPTER 5 – WHAT IS AN SCN? 31

Can you tell me a little bit about a DID? 32
What the PRI stands for - Primary Rate Interface 33
And I guess the SCN; it allows me to have my own number. 33
It allows me to keep my current phone number? 34
SIP stands for session initiation protocol. 35
How does that relate to the VoIP? 36

CHAPTER 6 – INFRASTRUCTURE CONSIDERATIONS 38

Inside the infrastructure, will my existing wiring work? 38
So, no need for additional LAN equipment or yeah? 39
This is telling me that all internet connections are not the same, I'm I right in assuming that? 40
So, how much bandwidth is enough? 42

CHAPTER 7 – HOW MUCH DOES A MODERN PHONE SYSTEM COST? 44

Is it going to be per day, per line, per hour? What are we looking at in price breakdown? 44
How would I know my current telephone system's worth? 45
Can you give me tips on just some basic systems? 47
How much maintenance, so what maintenance should be included to be included should I source out for that? ...48
So you mentioned leases, that stuck out to me. Do they

WORK? 49
COSTS, BUDGET, ALL MONEY TALK. WHAT'S YOUR ADVICE ON
BUDGETING? 50
HOW NOT TO SPEND MORE THAN WHAT I WOULD NEED...............51

CHAPTER 8 - COOL SYSTEM AND PHONE FEATURES.
....53

HOW ARE NOTIFICATIONS ABOUT VOICE MAIL DELIVERED?...........54
WHAT HAPPENS WHEN POWER IS LOST? 56
ANOTHER FEATURE THAT'S AVAILABLE IS-- OF COURSE, YOU'VE
GOT VIDEO CHAT. 57
WHAT CAN I DO TO BE MORE MOBILE? 58

CHAPTER 9 – THE INSTALLATION PROCESS 61

WHAT IS THE INSTALLATION PROCESS LIKE?...................................61
HOW LONG DOES INSTALLATION TAKE? 64
IS THERE DOWNTIME WHENEVER THERE'S AN INSTALLATION? 64
WHO PROGRAMS THE SYSTEM? 66
HOW IS TRAINING DONE? 68
WHAT IF IT ISN'T QUITE PROGRAMMED THE WAY THAT THEY
WOULD LIKE, AFTER THE INSTALLATION? 69

THE ULTIMATE GUIDE TO LAUNCHING A MODERN BUSINESS PHONE SYSTEM

INTRODUCTION

I have been in the telecommunications business since 1998, selling phone systems, phone service. And just to tell you a little why I'm doing this, I started working for a large manufacturer of telephone equipment, and then I ended up getting recruited to go work for a telephone company. And after a lot of downsizings, bankruptcies, and mergers, I had to come and make a decision on whether or not I was going to continue with this profession or move to something else.

And I came to the conclusion that I was better off just starting my own business and taking care of my customers, instead of continuing to bounce from one company to the next, abandoning my customers, where I could treat them as friends and handle issues that would come up and have a little bit more of an insight on some of the things that might go wrong and be able to fix problems and provide insightful solutions into the businesses, and how they wanted to do their communications and phone systems and that kind of thing. So, that's why we're here, and we're going to cover some of the things that you need to look for.

THE ULTIMATE GUIDE TO LAUNCHING A MODERN BUSINESS PHONE SYSTEM

CHAPTER 1 - Biggest Mistakes Launching a New Modern System

In this chapter, I'll reveal some of the biggest mistakes when launching a new modern business phone system.

What's the biggest mistake your prospects make?

One of the biggest mistakes that folks will make is getting sold versus buying what they want. And a lot of times what will happen is that a salesperson will come in, and they will have an agenda on a piece equipment or a platform that they want to sell, and it all sounds good whenever you evaluate and take a look at it, but at the end of the day, they didn't really take advantage of some of the other things that may've been out there, and they were only given one perspective and not necessarily buying a system that they want to get-- that's suited for them.

Not asking enough questions ahead of time.

And that goes, for not only you as the business owner, but also as the salesperson or consultant. Overlooking some things that could be addressed and by not asking some of the questions as the business owner that you want to have handled, you forego a possible solution that could have been implemented had you had the foresight or had the consultant, had the foresight to go ahead and take advantage of for your business, which leads us to the big picture.

Not taking a look at your entire platform.

What I mean by that, it's not just your phones, but the services connecting the phones that even on top of that, maybe even your internet connections and that kind of thing. If you're not taking a look at the whole architecture as a whole, you could be missing out on some solutions that are even better in the way that they would affect your business.

Not Getting Easy To Use Phones.

In some of the phone systems that are out there, there's a lot of feature codes and buttons that aren't able to be programmed. Either that or there aren't many hard-programmed buttons on particular phone models that allow you to take advantage of simple features. And so, if you're, again, just looking into a single phone system offered by some sales person that's just looking to make a quick sale and get down the road, you may not be able to easily use and take full advantage of the systems that are in place, and

THE ULTIMATE GUIDE TO LAUNCHING A MODERN BUSINESS PHONE SYSTEM

take advantage of all the features and functionality of a system that you install.

Got it. So that "ease of use" is knowing the full knowledge of how to utilize a phone.

Not utilizing the full capability of the system.

If you're having a hard time using features, you're not going to fully utilize the system. If you don't know how to do simple functions, you're more apt to just go ahead and skip over it and go back to using things the hard way in which, basically, eliminates the purpose of putting in a phone system, to begin with.

In this chapter, you learned the biggest mistakes in launching a modern business phone system. In the next chapter, I'll reveal the Telecommunications Spectrum and the best system.

THE ULTIMATE GUIDE TO LAUNCHING A MODERN BUSINESS PHONE SYSTEM

CHAPTER 2 – THE TELECOMMUNICATIONS SPECTRUM AND THE BEST SYSTEM.

Great opportunity to tell you about something that I came up with and that's what I call the business telecommunications spectrum, and it helps people identify what is the best system for them because there is no single best system out there. So, if you take a look at the graph that's on this page here, you've got a horizontal line that represents the actual service that your system will connect to and then the vertical line is going to represent the type of system, whether it be a digital system or an IP system. With the service, you can have analog service and that would encompass not only just regular, plain, old telephone lines or POTS lines as we say in the profession, but what is also known as a PRI or primary rate interface. What that is-- I will cover that here in a little bit-- but that also encompasses, I'll just say an analog signal versus SIP telephone service, which is an IP-based telephone service. Which, again, we'll cover

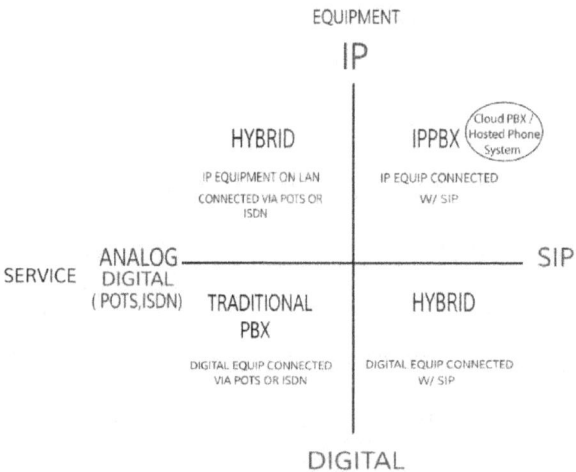

that here in a little bit but as you progress from

THE ULTIMATE GUIDE TO LAUNCHING A MODERN BUSINESS PHONE SYSTEM

the left to the right on the horizontal graph, you migrate from traditional telephone service to the IP telephone service, and the same thing goes in the vertical graph, from the digital on the bottom to the IP on the top. In the top, left-hand quadrant, you're going to have a-- why don't I just start with the bottom, left-hand corner which is the traditional phone system where you're going to have traditional telephone lines and traditional digital telephones on your desk.

As you move up to the left hand, you have a combination of IP system that's on the LAN that where the phones are connected in through your computer network and traditional telephone service that gives you the dial tone to the outside world. Once you move to the right side of that vertical line, you move into what I call another different type of a hybrid system where you still have digital phones, but now you're using the internet connection to actually make and receive telephone calls. That's another different type of a hybrid system. Then, we move up to the top, right-hand quadrant; we're talking about having an internet connection to make and receive calls, and the phones are on your LAN. We call that an IP PBX. Internet protocol PBX and then the latest evolution in all that is where you take the IP PBX which is the premise equipment that is in your phone closet, if you will, and then they move that out into the cloud, where that piece of equipment is now cloud-based, and everything is accessed through the Internet. Every call that you make traverses the Internet and comes back. It originates in the Internet and comes into your desk, via the Internet. That's the latest evolution in the

THE ULTIMATE GUIDE TO LAUNCHING A
MODERN BUSINESS PHONE SYSTEM

telecommunications spectrum.

In this chapter, you learned the Telecommunications Spectrum. In the next chapter, you'll discover call handling philosophies and strategies.

CHAPTER 3 – CALL HANDLING PHILOSOPHIES AND STRATEGIES.

In this chapter, I'll reveal call handling philosophies and strategies.

There's a certain myth about voice mail and auto-attendance, sometimes people think, well I don't want voice mail. Well, a lot of the times, once I clarify what they're talking about there and not necessarily wanting to eliminate voice mail or not get voice mail, but what they're really talking about is the utilization of what we'll call an auto-attendant, where whenever someone calls into your business, an automated attendant picks up the call and handles the call immediately versus a live answer. So, this is one

THE ULTIMATE GUIDE TO LAUNCHING A MODERN BUSINESS PHONE SYSTEM

of the philosophies when it comes down to the strategizing and how you want to handle calls in your business is, do you want a live person to answer a call, do you want that personal touch, or you more or less want to get your work done and are hands-off and expect the automation of the system, and the purpose of it in the first place, is to lighten the workload of a receptionist or a live attendant to handle each and every call. So, these are some strategies that you just kind of have to evaluate depending on your personal philosophy and how you run and operate your business.

It seems that somebody would be-- next question they would probably ask is

How could they screen their

David Bridge

calls?

That's a great question, and one of the initial ways to do it would be through the implementation of an auto-attendant, in where, again, you just have the automated attendant pick up the call, it captures the [chuckles], yeah, it captures the caller ID information from each and every caller, and then that transfers based on their selections within sub-menus and transfers the calls based on their selection. If they make no selection or hit the zero, it's going to go to their section store operator.

And is that automated attendant, I guess can it detect who someone is unsure about speaking about at the moment or--

Yes, so this relates to a specific feature on some phone systems. Depending on the system, I'll

THE ULTIMATE GUIDE TO LAUNCHING A MODERN BUSINESS PHONE SYSTEM

just simply refer to it as your old answering machine days, where you wait until it picks up, and if you're a fan of Seinfeld [chuckles], I'll start singing the song, "I'm not at home." And then you just wait for the beep and wait till they're starting to leave you a message. And then at that point, you can determine, "Oh, he's calling about football tickets. I better take that call," or if he's just calling to beat your ear about the beating that the football team got the night before, whatever. Then you could just make that determination. But some phone systems have that kind of capability into it.

Does that bump out the receptionists?

No, it wouldn't necessarily bump out the receptionists because if a particular caller wants to zero-out, as we say, then they can go to and get a live answer where if they don't necessarily know who they want to speak to, the receptionists then us on her set based on that zero-out can determine that the caller actually did that versus someone that's actually dialing her particular extension, she'll have that notification via a display on her phone.

What's the biggest mistake a receptionist makes screening calls?

Sometimes they won't know, and the system won't have been programmed based on how they wanted it to be handled, and so if the 0 is 0, and it doesn't differentiate anything along those lines, whether it was a zero-out or it's being called for an extension, she won't particularly know how to answer a call in that way. But another one is, someone's personal philosophy on actually taking calls. Let's say that it's the president of the company that has a philosophy of answering every call no matter what type of work he's in the middle of, even if he's working on something that's important and he'd be better off hitting the "do not disturb" on his phone and foregoing that conversation and letting it go to voicemail as opposed to just picking up and answering it and then trying to get back into the swing of things of what he was doing before. But the biggest mistake is not figuring out and having a specific philosophy on when and where and how you're going to actually receive your personal phone calls.

Another call-handling strategy question is; how

David Bridge

do you handle calls for a specific department like sales or customer support?

Well, when a call gets added to a department like sales or customer support, there's a few different options one is you can have every phone in the department ring at the same time, and then whoever's available can pick it up. Another way is uniform call distribution, where it just rings from one station to the next, rotating in an order that it was already received. And then another one is in order, so that at each time it starts at the first extension, if they're busy, goes to the second. The difference being, in the uniform call distribution, if the last call is at the second-- if it was taken by the second, the next call would be rodded to the third extension in

THE ULTIMATE GUIDE TO LAUNCHING A MODERN BUSINESS PHONE SYSTEM

the rotation, where in order it'll start at the beginning every time. And then of course you've got automatic call distribution, where you've got queues that can be built so that if four calls come in and only three people can handle calls, the fourth call will then be put into a queue and then the first available agent will then answer that call as soon as they're done.

In this chapter, you learned call handling philosophies and strategies. In the next chapter, you'll see how to connect multiple sites.

David Bridge

CHAPTER 4 – CONNECTING MULTIPLE LOCATIONS

In this chapter, you'll get to see how to connect multiple sites.

As businesses grow, they'll typically have more than one location, and so I'll get the question and ask how can we connect multiple sites? And there are a number of different ways. One is through the IP technology and what we'll do is we'll set up a connection that allows interoffice communication through the internet.

So I guess it doesn't matter how far away they are?

The distance isn't as relative, or relevant I should say, as it is specifically where the location is. If a particular location is in the middle of nowhere and they don't really have a great internet connection where they can take advantage of the IP telephony to the connect a couple of different sites. Then what ends up happening is you can perform a what we'll call a Trunk-to-Trunk Transfer, where it's going to actually take up two lines to transfer a call, if you will, to a different site. But if another location doesn't have a very good connection to the internet it is going to be tough to connect sites. The distance isn't as relative as actually where the location is.

David Bridge

So should I improve my internet connection?

There are tests. It really depends, there's test that you can perform ahead of time. And then you can figure out what would you need to do-- improve the connection or sometimes what is available. And doing that is implementing what is known as a direct internet connection. Some service providers out there can provide a direct ethernet connection from one site to the next via fiber connections or cable connections. And with that in place, you can implement the IP telephony that we were talking about earlier.

What if one person works in more than one location?

So if someone specifically, one particular person, wants to work in more than one location they've got a few options. You can take an IP phone with you as you move from site to site. What that'll do is, it will register on the LAN as your phone. And you can take and make and receive calls, and it doesn't matter where you've set the phone up. It doesn't even matter if you're in one of your office sites if you've set it up correctly. You can take it and plug it in in China and appear as though you're on site. In addition to that, if you want, you can actually leave the phone and go to a different desk at the other location and have that phone mirror the way you do things at the first location. And that can retrieve messages, and they'll be deleted off the first phone. Another option too is if you're going to employee the cloud PBX that we were talking about earlier, if you're that level, maybe you just take advantage of some of the mobile applications. That might be the way to go where you just have the mobile app, and you use that as your phone.

How do I keep from tying up multiple lines connecting another location?

That's one of the ways that if you're going to do-- you simply can't do it if you're going to do a digital system. If you're going to transfer a call from one site to the other and you're not employing IP telephony, then you're just going

THE ULTIMATE GUIDE TO LAUNCHING A MODERN BUSINESS PHONE SYSTEM

to be tying up the line to come in and the line to go out using the trunk-to-trunk transfer. But if you got IP telephony in place, then what you can do is set that up so that it's just a matter of transferring a call or intercoming a call. And then the same applies to a hosted system, too, where there are no control units to install - the control unit is in the cloud - and any and every phone is tied back to that. So wherever you place the phone, it's as though you're all in one site.

Will I know they're on the phone?

One of the buttons, if you want to program it as a specific busy lamp field-- if the person is on the phone it will light up and appear a different color as opposed to do not disturb. And you'll be able to tell based on that busy lamp field, whether it's lit or not, whether they're on the phone in the office or whatever.

What if I want an intercom in my offsite office?

Okay, then you just enter the feature key, if you've got it, depending on the system, you enter the code for that particular office and their extension. Or you just program a button on your phone to intercom a particular department, station, or just the receptionist.

Switch gears a little bit.

THE ULTIMATE GUIDE TO LAUNCHING A MODERN BUSINESS PHONE SYSTEM

What is fiber ethernet?

So this may be a little bit premature to talk about this, but it kind of goes back to if you're going to connect sites, then fiber may be a connection that you want to take advantage of in that it's a real pure, synchronous connection to provide reliable internet access. So that if you're going to use IP telephony, it may be something you want to look in to.

Can set it up to ring a different site, program it to ring a different site?

That's going to be part of an automated attendant. If you want to program in a sub-menu to say that something like, "Hey if you want to file a claim with ABC company," or something like that. Or, "If you want to reach our other office," you can program it in as an out-dial if it's a digital system. Or you can program it essentially the same way whether it's a hosted system or an IP system.

In this chapter, you saw how to connect multiple sites. In the next chapter, I'll reveal what an SCN is.

THE ULTIMATE GUIDE TO LAUNCHING A MODERN BUSINESS PHONE SYSTEM

CHAPTER 5 – WHAT IS AN SCN?

In this chapter, we're going to talk about an SCN - single communication number.

And what that will allow us to do is, just to give one particular user, their own personal telephone number. And it's not their cell phone number; it's their own business number that's identified with the business. So that if an employee leaves, that number stays with the business and then the business owner doesn't have to worry about, "Well, geez, all the calls we were forwarding them to their cell phone or the business card has all the employees cell phone numbers on it." We can still capture the calls from the sales people who are still receiving, you know, that there're business cards that's still out there. We can still capture those sales inquiries on calls made to that single communication number.

Can you tell me a little bit about a DID?

So the DID kind of ties in with the single communication number in that it allows individual numbers. It can be done through the hosted platform where a DID is assigned. PRI can also have a block of DID numbers which the next question is the PRI.

THE ULTIMATE GUIDE TO LAUNCHING A MODERN BUSINESS PHONE SYSTEM

What the PRI stands for - Primary Rate Interface

It's a service delivered that can be a fractional PRI or for a full PRI and it usually starts from anywhere from 8 to 10 channels up to 23 channels where it varies depending on the caller traffic. So we don't have to designate 3 or 4 lines for outgoing traffic, 4 or 5 lines for only incoming traffic. Any particular trunk can be an incoming call, an outgoing call; it can also be a call associated with a specific number that's not an actual physical number which is what a DID is what we were talking about before.

And I guess the SCN; it allows me to have my own

David Bridge

number.

That's exactly right. So whenever you have a DID or an SCN - is what we call it - you can have your own individual number, and it's not this person's cell phone number, it's a number that's associated with the business.

It allows me to keep my current phone number?

And what that also will allow you to do, regardless of whatever system that you implement is, your main business telephone number is what we'll call portable you could take that main number 123-1200, regardless of what service or provider you use, you'll be able to move that number from one provider to the next, you don't have to worry about advertising or business cards, letterhead and that kind of thing. So you can keep and utilize that same number over and over again.

Yeah and here's another fun acronym as it related to the DIDs and the PRI, sip or SIP.

> SIP stands for session initiation protocol.

That's a service that is strictly an internet based service. And it can be - you can deliver DIDs over SIP. You're going to need an internet connection for SIP. And the funny thing is it's a -

as a back story, a lot of the infrastructure calls the telephone companies use themselves are going to be based on the SIP protocol, the session initiation protocol. It's hard to find a phone call today that somewhere along the way it wasn't a SIP phone call.

How does that relate to the VoIP?

VoIPs. Voice over Internet protocol. So, voice over internet protocol. SIP is part of the voice over internet protocol, and some people will get kind of confused on that term in that they can have IP phones on their premises and those are certainly voice over internet protocol phones, but then there's also the differences between that and making a phone call to the outside world using SIP protocol, which is typically going from the customer site out to the public.

We talked about a lot of different systems and services, even gone into detail with DID, SIP, and VoIP.

In this chapter, you discovered what an SCN is. In the next chapter we discuss infrastructure considerations.

David Bridge

CHAPTER 6 – INFRASTRUCTURE CONSIDERATIONS

In this chapter, we'll discuss infrastructure considerations.

Inside the infrastructure, will my existing wiring work?

If you're going to install a new digital system, chances are that's what you're replacing. So there aren't going to be much worry in the way of rewiring your infrastructure for digital phones. However, if you're going to go from a digital system to an IP-based system, you want to take a look and make sure you have-- you've got at least CAT-5E wiring in place that are connecting all of your computers to your router. In addition to that, you want to make sure that wherever you're putting your phone that doesn't have a computer, you wire up that location too. A lot of the phones that are IP phones have a port in the back of them so that you don't have to add an additional data line then that you can you use the existing data line to unplug your computer from the LAN plug that connection into the phone and then from the phone back into the computer and you can still utilize it in that way.

So, no need for additional

David Bridge

LAN equipment or yeah?

So it depends on whether or not you've got enough ports on your particular router, whether or not you need to add anything, if you're going to take advantage of any IP-based phone system. You may need to upgrade it and look at whether or not some quality of service issues are adjustable too-- you know there are tests that you can perform on your network to figure out and make sure that if you're going to do voice over IP on site, that it's going to be able to handle that additional load.

This is telling me that all internet connections are not the same, I'm I right

THE ULTIMATE GUIDE TO LAUNCHING A MODERN BUSINESS PHONE SYSTEM

in assuming that?

Yeah. All of the internet connections are not the same; it's hard to find anybody that's choosing DSL anymore, but they're out there. That's the lowest internet speed that I've found these days. If you're using that, I would strongly encourage you to consider changing that to something on the order of cable or even a fiber connection depending on the scale of the business that you're going to-- that you're going to operate because a digital-- a DSL service - I'll just stick with the name of it - but a DSL service is probably the lowest in that you want to have it to access the internet and take advantage of SIP calls, hosted calls and those forms of communications. That can't affect you so much if you're just going to have an IP-based phone system that's on your LAN that's still you using traditional telephony. But if you're going to do anything off-site, connect anything in the way through the internet to another location or use SIP trunks then you definitely want to consider something else.

So, how much bandwidth is enough?

That's going to depend on the number of calls that you're making. Again, you want to perform some tests that whoever you're looking at should be able to direct you to-- there are some simple ones online if you just type in 'simple tests' like in the Google VoIP test, you'll get directed to a number of different sites. But you're a vendor if you're going to-- seriously

THE ULTIMATE GUIDE TO LAUNCHING A MODERN BUSINESS PHONE SYSTEM

considering according to any phone system should be able to perform some kind of test to let you know whether or not your network is stable enough to handle a voice over IP system.

In this chapter, we discussed infrastructure considerations. In the next chapter, you'll know how much modern systems cost.

David Bridge

CHAPTER 7 – HOW MUCH DOES A MODERN PHONE SYSTEM COST?

In this chapter, you'll know how much is this going to cost?

Is it going to be per day, per line, per hour? What are we looking at in price breakdown?

There are a number of different components to consider when it comes to the actual cost of a phone system. The biggest consideration is the

THE ULTIMATE GUIDE TO LAUNCHING A MODERN BUSINESS PHONE SYSTEM

number of phones that you're going to have. If you're going to install any new circuitry to make phone calls, that's going to be another consideration. You can typically look on the low end, $200, $300 dollars per phone. On the high end, you could spend upwards of $800, $900 per phone. If you really want something, that's going to handle a significant requirement in terms of telephony. In terms of service, you can get SIP trunks as low as $12, $13 per line and if you're going to go hosted service you can expect to spend $15, $20 per phone up to $35, $40 per station, per seat, per license if you will.

How would I know my current telephone system's worth?

That's a good question. There's a number of different ways sometimes - the provider that you're looking at may have a trade-in program. That's a question worth asking. Sometimes you can get a couple of dollars if you throw some phones up on eBay. If you're a business owner it may--, it's going to depend on how frugal you are, on whether or not you actually want to take time to list a phone system up on eBay. If you've got some kids maybe throw it in their lap and let them do it. Sometimes there are wholesalers out there looking for used equipment to market to third world countries too that you may be able to find and just-- they'll give you a decent offer on some of their equipment that you're looking to replace.

I don't need anything fancy.

THE ULTIMATE GUIDE TO LAUNCHING A MODERN BUSINESS PHONE SYSTEM

Can you give me tips on just some basic systems?

So, it's not really about fancy; it's about your philosophy. Whenever you have a discussion with a proposed vendor, you want to make sure that whatever you're installing meets the way you do business. If you're not into technology, just make sure you let them know that. If you're wanting to take advantage of all the things that systems these days are capable of, then that's a whole other rabbit hole to go down.

Now, can you give me some advice on comparison? As if I like my current provider and I feel no need to upgrade, or--

You know what, I tell anybody if they like who they're dealing with at present, I just tell them to go ahead and keep them. If you like them, there's obviously a reason that you like them. Not a good-- I'm of the philosophy that loyalty is a good thing.

How much maintenance, so what maintenance should be included to be included should I source out for that?

That's a good question to ask the vendor you're looking at. It's up to you. It could be included in a lease if you're going to purchase a premise based system and then you can renew annual agreements as you see fit. A lot of the times people that come and have a high turnover, it's worth looking into. If you've got a stable business with people who have been there a significant amount of time, it might not be something that you're going to move phones around, especially too if you're not really growing, it might not be something that you feel you can just pay it, time and materials.

So you mentioned leases, that stuck out to me. Do they work?

Yeah, of course. It's one of the things that you will want to address with your account. That's a question that best asks them, but they do allow you to upgrade without putting a big cash outlay. If that's something you're considering, you may even want to consider a hosted platform where you'll pay depending on the provider, an installation cost, sometimes you can get away with installing the systems yourself, if you feel comfortable with that, then pay subscription on the hosted platform with the small initial upfront and installation. Or I'll say nominal installation costs for a hosted system.

Costs, budget, all money talk. What's your advice on budgeting?

Someone that has a philosophy of they're not just going to sell you a phone system, but are going to take a look at your entire communications platform, "From soup to nuts" as they say, may be able to find ways to, not only improve the way you're communicating

THE ULTIMATE GUIDE TO LAUNCHING A MODERN BUSINESS PHONE SYSTEM

and offset the cost of a system through what we'll call soft cost, through improved efficiently and productivity, but they may be able to introduce services that will offset the cost of the system on top of that.

How not to spend more than what I would need.

Yes. So whenever you're doing that, taking a holistic look at your platform, even going out-- sometimes people go outside your comfort zone a little bit, but with some of the enhancements of the hosted platforms these days, with the improved call quality, those are some things you may not have to even worry about busting the budget worth. Sometimes phones can be included in the service; again you're only paying an installation cost.

From this chapter you now know how much a modern phone system costs. The next chapter reveals some cool system and phone features.

THE ULTIMATE GUIDE TO LAUNCHING A MODERN BUSINESS PHONE SYSTEM

CHAPTER 8 - COOL SYSTEM AND PHONE FEATURES.

In this chapter, you'll hear about some cool system and phone features.

These days, there are some touchscreen phones that are available for your desktop, and it just allows you to treat it like an iPhone or an Android. Touch the screen, and you're activating the buttons as you would as though they were hard key buttons. Another thing about some of the phones that are in addition to the touchscreen that isn't necessarily touchscreen phones but they are labelless phones; there's no paper necessary. There're LCD buttons that are programmable so that if a button gets updated, you don't have to rip the paper out and get the white-out out and redo the paper label. The buttons are labeled through an LCD, so as you program-- do it-- update the label in the programming, the LCD just updates automatically.

How are notifications about voice mail delivered?

So you've got your standard way of being notified about a voice mail in that there's a button on the phone that will light up. You can have-- if you've got the right system, have a voicemail sent via email as an attachment, as a

THE ULTIMATE GUIDE TO LAUNCHING A MODERN BUSINESS PHONE SYSTEM

WAV file, or as a link. If you've got a hosted system, apps can be updated so that you can check your voice mail through the app on your smartphone, and then that handling of the voicemail, whether it's saved, deleted, you know marked as not new anymore but is listened to, will update the entire system. One of the other ways is that you can have--once a voicemail is left on the system, you can have the system then make an out call if you don't have a hosted system or an IP system to your phone to let you know that there is a message in your voicemail box. Then of course if that-- you know some of the other old ways would be a stutter dial tone. If you're getting a stutter dial tone in your system, you know that you've got voicemail through the phone company on that particular line that you need to make a phone call to-- or a code call to retrieve that message. So those are some of the ways to know about notifications for voicemail.

What happens when power is lost?

This is a neat feature for some of the hosted systems in that if power is lost to the building or a particular station, pre-programmed numbers and devices can be set so that once a cloud system knows that a station or the entire system has gone offline, it knows to route a particular-- your call or office calls to designated numbers ahead of time that are in the system, so you're not going to miss a call and you don't have to worry about losing power, losing connectivity, and you've got business continuity built into a hosted system. Now if you've got a battery back-up for a premise based system, that's the best way to ensure power loss, but if you're using SIP trunks on a premise based system and you lose internet connectivity, same thing will apply, you'll have pre-programmed destinations in the cloud, if you will. If you're using SIP trunks, if your Internet connectivity goes off, calls will get routed to specific numbers pre-programmed ahead of time.

> Another feature that's available is-- of course, you've got video chat.

Those are available through FaceTime and Google Hangouts, but some phones and some hosted systems will have the ability to enable you to do some video chat. Another replication of text messages is the ability to chat via a chat box that replicates text messages. This can be done through the system itself or when a station comes online, or a user logs into their dashboard. And you know that they're in the office, and you can initiate a chat sequence without having to actually make a phone call. Another way around this too is to use Google Hangouts. Any user that's logged into their Hangout or via Gmail can initiate just an online chat between users.

What can I do to be more mobile?

Well, this goes to the hosted platforms in that

THE ULTIMATE GUIDE TO LAUNCHING A MODERN BUSINESS PHONE SYSTEM

you can have an app that replicates your station, and when you make or take a call via the app, it appears as though you are on site or you're logged into the business to make and receive those calls. And, again, you can check voicemails and intercom a station and appear as though you're on site or essentially logged into your work or business what happens if you lose Internet connectivity is another question, but then that goes to one of the features of the business continuity. So that whenever you've got Internet-based services like SIP trunks or hosted phone system, preprogrammed numbers are already in the system, already programmed in your cloud system. So that when it notices a station or a line goes off, then calls are automatically rerouted to a cell phone, a different landline so that business continuity is maintained.

David Bridge

In this chapter, you discovered some cool system and phone features. Installation matters are what the next chapter is about.

THE ULTIMATE GUIDE TO LAUNCHING A MODERN BUSINESS PHONE SYSTEM

CHAPTER 9 – THE INSTALLATION PROCESS

In this chapter, you'll learn what the installation process is like.

What is the installation process like?

What will happen is we'll ask a lot of questions prior to the installation. Namely, the names of the stations that are going to be put in place or replaced. How many lines are in place? How calls that come in are to be routed whether they go to departments whether they go to individual people? Are there DIDs, direct inward dial lines, that need to be programmed or replaced, and if so, which station do they go to? How are voicemails going to be setup? Those are some simple questions and then everything else is pretty much tailor-made to the individual business on the questions that we'll ask in how calls will be routed and how calls will be made, fax machines, and that kind of thing, anything else to be made aware of. Also, if there's going to be a hosted system put in place, a lot of times the phones can be put down right alongside the existing phones because everything is based on the Internet and you don't need to necessarily replace a phone, a digital phone, and swap it out without having to go without using a phone. That can be a little smoother in that there's a little less disruption, but there is an extra phone at the same time. As the phones get placed, they can be programmed.

THE ULTIMATE GUIDE TO LAUNCHING A MODERN BUSINESS PHONE SYSTEM

During a hosted installation, it's basically a matter of the numbers being turned up and made live or ported - in other words; they are transferred from one carrier to the next. If additional services are going to be installed - in other words, if you're going from a regular POTS lines to an ISDN service - of course, you're going to have to make sure the equipment is in place to handle that transition ahead of time. Otherwise, the timing of the installation doesn't really matter if you're switching carriers but not switching the way the service is delivered. Either the timing isn't quite as critical. If you're switching how the service is delivered, like you're going from POTS lines to SIP trunks or POTS lines to an ISDN, then, of course, you want to have the equipment there that's going to be able to handle that transitional phase.

David Bridge

How long does installation take?

The actual installation has got to be done, but there are pre-installation questions and work that needs to be done in that you just need to know where the phones are, what the environment is going to be like, if it's a premise based system is there room, is there power to connect the system or to hang the system in the same room as an existing system? But that's usually going to take and be done within one day. Of course, you've got programming that's done ahead of time, and sometimes even some adjustments to the programming after the fact.

Is there downtime whenever

THE ULTIMATE GUIDE TO LAUNCHING A MODERN BUSINESS PHONE SYSTEM

there's an installation?

There shouldn't be any downtime, unless there's something that really goes wrong, but the way the systems are installed you can avoid any downtime by putting some phones in place so that you can do some tests and then put in phones that are not quite being used as much and doing some more tests, and then gradually work your way up and doing a cutover, as service in the phones are being cut over, but anybody that knows what they're doing doesn't have to-- you know they can work around and make sure there's a minimal loss of downtime. Whenever you've got premise based phones they've got-- the phone itself will be on the desk, but not live, at the same time, unless you're moving from a digital to an IP-based system, but if you're going from a digital to a digital the phone will just be there, probably at the same time, but not quite live, because the actual wiring just needs to be switched over to the system, but if it's-- when it's done right it doesn't take long at all, you'll lose very minimal downtime, but again if it's a hosted system, then it's just a matter of waiting on the actual port to be done from the old carrier to the hosted platform platform, and that's just a matter of a couple of keystrokes, so when the phones are all ready and in place, and the port is done, the phones are live, and calls are just automatically-- it's such a seamless transition, and there is no downtime.

Who programs the system?

THE ULTIMATE GUIDE TO LAUNCHING A MODERN BUSINESS PHONE SYSTEM

Another question that comes up, usually there's a project manager that will program the system ahead of time and is going to be needed-- probably some adjustments that will need to be made after the installation just to make sure it's precisely the way the user wants. Whenever it's a premise-based system, you know you've usually got just admin ability that's handed over, but of course, when the user buys the system they've got-- they own the system, so there isn't any hesitation to hand over the full access to the programming assistant.

A lot of times there are things that can be goofed up by giving full access unless there's someone that's fully trained within the business to do the actual programming. And with hosted systems you've got a dashboard that you can fool around with the user, fool around with the programming, you know the entire platform and program it just as-- just the way you like if you're the user, or you can you know, not have to worry about it and have your professional communications vendor take care of the programming for you, so you don't have to sweat it, but you're given full access either way.

How is training done?

One of the questions about training, the typical way to do training for a new business on the phone system is for the process of training the trainer to be done, unless you've got some kind of advanced concierge service that you subscribe to, where you know-- this is certainly available where you can have everyone be individually trained on how to use each of the phones and each of the platforms and all of that, but typically what happens is a project manager or a trainer will train someone on staff to then make sure everyone within the business knows how to use the new phones. Training the trainer is what we call it, whenever you're learning about a new system, but again, you have more advanced packages that will allow for training of each user to use their phone.

What if it isn't quite programmed the way that they would like, after the

David Bridge

installation?

A lot of times that's just going to be part of the installation, you get some kind of what we call the [debugging], you can just have the system programmed and tweaked to just the way you want it, but ongoing program can be part of a maintenance agreement, where you know, if you just want something done you can just make a phone call and have that taken care of, or if you feel comfortable with it you can certainly log in to the system and have the ability to program it and change the programming just the way you want

From this chapter, you now understand what the installation process is like.

THE ULTIMATE GUIDE TO LAUNCHING A MODERN BUSINESS PHONE SYSTEM

To Contact David Bridge at Virtual Communications LLC for Consulting, Speaking or Interviews go to VirtualCommLLC.com or call 814-317-6470

www.ingramcontent.com/pod-product-compliance
Lightning Source LLC
Chambersburg PA
CBHW060413190526
45169CB00002B/883